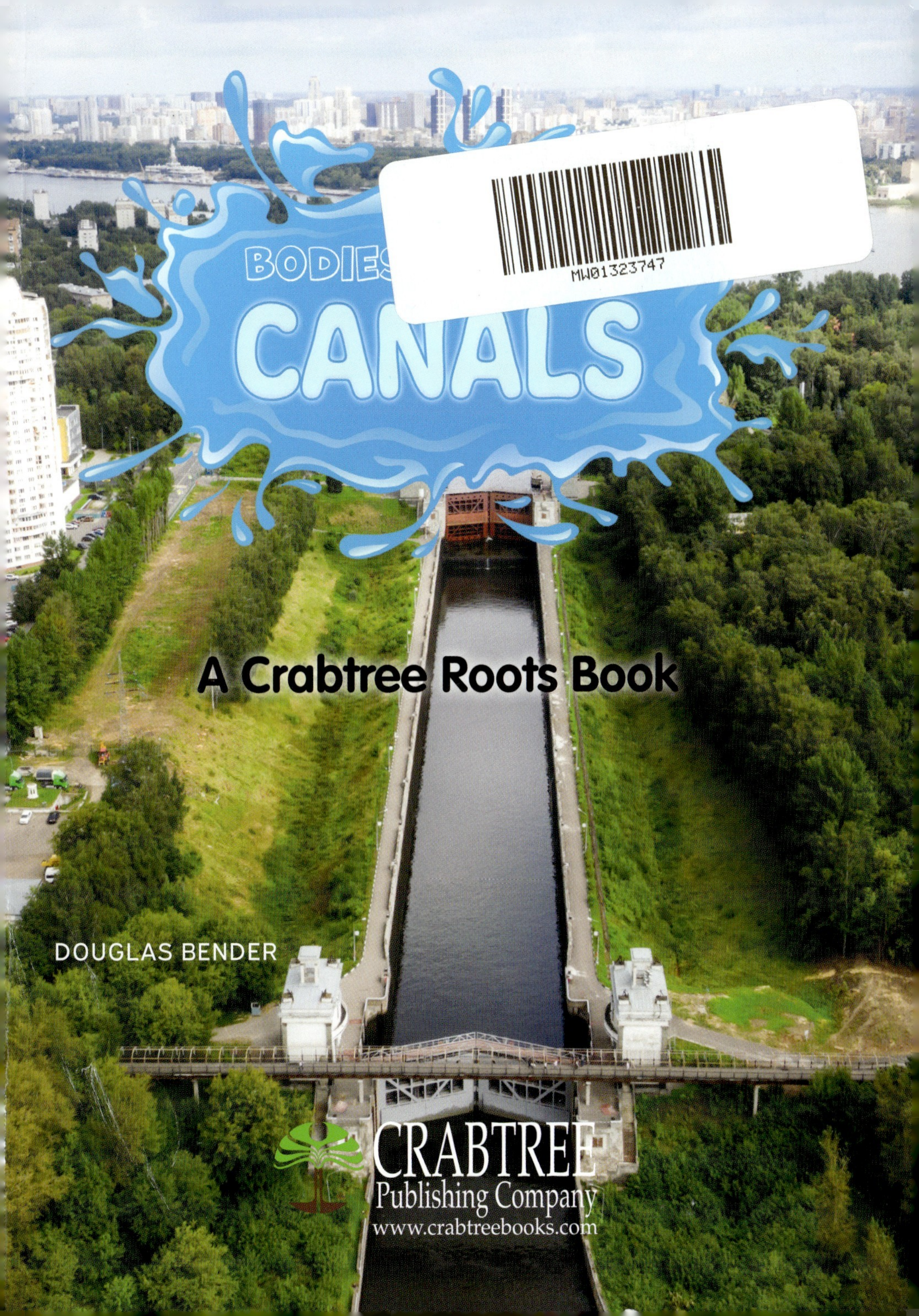

BODIES
CANALS
A Crabtree Roots Book
DOUGLAS BENDER
CRABTREE
Publishing Company
www.crabtreebooks.com

School-to-Home Support for Caregivers and Teachers

This book helps children grow by letting them practice reading. Here are a few guiding questions to help the reader with building his or her comprehension skills. Possible answers appear here in red.

Before Reading:

- What do I think this book is about?
    - *I think this book is about water and canals.*
    - *I think this book is about how people use canals.*
- What do I want to learn about this topic?
    - *I want to learn what a canal is.*
    - *I want to learn why people build canals.*

During Reading:

- I wonder why...
    - *I wonder why canals use dams.*
    - *I wonder why boats use canals.*
- What have I learned so far?
    - *I have learned that canals help move water to cities and farms.*
    - *I have learned that canals are made by people.*

After Reading:

- What details did I learn about this topic?
    - *I have learned that canals help connect waterways.*
    - *I have learned that some cities are built far away from water.*
- Read the book again and look for the vocabulary words.
    - *I see the word **canal** on page 3 and the word **dams** on page 10. The other vocabulary word is found on page 14.*

This is a **canal**.

Canals are made
by people.

Many canals help
get water to **cities**.

Some canals help get water to farms.

Some canals
use **dams**.

Most canals can be used by boats!

Word List

Sight Words

a
are
be
boats
by
can
farms
get
help
is
made
many
some
this
to
use
water

Words to Know

canal

cities

dams

34 Words

This is a **canal**.

Canals are made by people.

Many canals help get water to **cities**.

Some canals help get water to farms.

Some canals use **dams**.

Most canals can be used by boats!

Written by: Douglas Bender
Designed by: Rhea Wallace
Series Development: James Earley
Proofreader: Janine Deschenes
Educational Consultant: Marie Lemke M.Ed.

Photographs:
Shutterstock: Solarisys: cover; Mikhail Starodubuv: p. 1; tetiana_u: p. 3, 14; Chockdee Permploysiri: p. 5, 14; RatFace: p. 6; Anton Havelaar: p. 8-9; Frank Legros: p. 11, 14; JaySi: p. 13

Library and Archives
Canada Cataloguing in Publication

Title: Canals / Douglas Bender.
Names: Bender, Douglas, 1992- author.
Description: Series statement: Bodies of water | "A Crabtree roots book".
Identifiers: Canadiana (print) 20210190108 | Canadiana (ebook) 20210190140 | ISBN 9781427155900 (hardcover) | ISBN 9781427155962 (softcover) | ISBN 9781427133809 (HTML) | ISBN 9781427134400 (EPUB) | ISBN 9781427156143 (read-along ebook)
Subjects: LCSH: Canals—Juvenile literature.
Classification: LCC HE526 .B46 2022 | DDC j386/.4—dc23

Library of Congress
Cataloging-in-Publication Data

Names: Bender, Douglas, 1992- author]
Title: Canals / Douglas Bender.
Description: New York, NY : Crabtree Publishing, [2022] | Series: Bodies of water - a crabtree roots book | Includes index.
Identifiers: LCCN 2021017146 (print) | LCCN 2021017147 (ebook) ISBN 9781427155900 (hardcover) | ISBN 9781427155962 (paperback) | ISBN 9781427133809 (ebook) | ISBN 9781427134400 (epub) | ISBN 9781427156143
Subjects: LCSH: Canals--Juvenile literature.
Classification: LCC TC745 .B46 2022 (print) | LCC TC745 (ebook) | DDC 627/.13--dc23
LC record available at https://lccn.loc.gov/2021017146
LC ebook record available at https://lccn.loc.gov/2021017147

Crabtree Publishing Company
www.crabtreebooks.com 1-800-387-7650 Printed in the U.S.A./062021/CG20210401

 In Canada: We acknowledge the financial support of the Government of Canada through the Canada Book Fund for our publishing activities.

Published in the United States
Crabtree Publishing
347 Fifth Avenue, Suite 1402-145
New York, NY, 10016

Published in Canada
Crabtree Publishing
616 Welland Ave.
St. Catharines, Ontario L2M 5V6